CARVIREN
Study Case

RealTrack Systems

Authors:
José Pino Ortega
Carlos Padilla Sorbas
Isabel Pérez Segura
Manuel Vera López
Victor Andrés Andrés
Miguel Ángel López Vicente
José López Cara

Illustrations and Cover:
Raúl Gómez Salvador

ISBN: 1503205827
ISBN-13: 978-1503205826

Special thanks to University of Southampton IT Innovation Centre, Institute of Communication and Computer Systems of the National Technical University of Athens, Atos, Joanneum Research, Bearing Point Infonova GMBH, Centre D'Alt Rendiment Esportiu de Sant Cugat del Valles, K U Leuven and The Interactive Institute Centre for Research and Technology Hellas

Index

Summary 7

1.   Introduction 9

2.   What is Experimedia 11

3.   RealTrack Systems 14

4.   Experiment Description 17

   4.1.   Sport Areas of Experimentation 18

   4.2.   Learning Objectives 18

   4.3.   Experiment Procedure 19

5.   Background of the Experiment 20

6.   RealTrack baseline component: WIMU 22

7.   EXPERIMEDIA Baseline Components 24

   7.1.   Experiment Content Component (ECC) 24

   7.2.   Audio Visual Content Component (AVCC) 25

   7.3.   Social Annotation Service 27

8.   REALTRACK Developed Components 31

   8.1.   WIMU: Mobile Transmission Unit 31

   8.2.   Sport Widgets (Sportwis) 32

      8.2.1.   Sport Widgets List 32

   8.3.   CARVIREN website 34

   8.4.   Experiment Areas 34

9.   Depth Review of the Trampolining Widget 38

10.   Depth Review of the Moxy Widget 45

11.   Experiment Architecture 51

   11.1. RealTrack Components 51

      11.1.1. Sensors 52

4

11.1.2. **DAPS** 52

11.1.3. Sport Widgets 53

11.1.4. CARVIREN Website 53

11.1.5. RealTrack Systems Component Architecture 54

12.   RealTrack Components and EXPERIMEDIA Components 55

12.1.   Functionality of the System 56

13.   Experiment Preparation 57

13.1.   Participants and Venue 57

13.2.   First Run 57

13.3.   Second Run 60

14.   Experiment Execution 65

14.1.   First Experiment Run 65

14.2.   Second Experiment Run 66

15.   *Quality of the Experiment Conclusions* 69

15.1.   Conclusions 70

16.   Ethics and Privacy on the Experiment 73

17.   Conclusions 74

17.1.   General 74

17.2.   Feedback on EXPERIMEDIA Baseline Components 75

17.2.1. *ECC Feedback* 75

17.2.2. AVCC Feedback 76

17.2.3. Social Annotation Service Feedback 76

18.   Experimedia Partners 78

18.1.   Core Partners 78

18.2.   Experiment Open Call 1 80

18.3.   Experimentes Open Call 2 81

Summary

This case of study reports on the design, development, implementation, execution and evaluation of the EXPERIMEDIA experiment for *CAR VIRtual ENvironment (CARVIREN)*, which focuses on a network for the CAR of Sant Cugat Venue; where information about the workouts of Biathlon and Trampolining are available to the "users" using a website.

The document is divided in three parts. The first part is about Experiment Description, and relates the whole background of the experiment, objectives and the baseline components provided by the Experimedia Consortium.

The second part involves the study of the RealTrack Systems' components. We will do a depth review of the new developments.

The third and final part regards the implementation of Experimedia and RealTrack components. We will check the runs that have been conducted at the CAR Venue and we will provide an evaluation of the Experiment.

Part 1:

Experiment

Description

1. Introduction

This case of study is split in two parts. The first part gives the background of the experiment: Experimedia description, Realtrack used components, Experimedia baseline components, main goals and work programme. The second part is about the integration, the development of the components (focused on the widgets), runs and results of the experiment.

CAR Virtual Environment (CARVIREN) is an experiment conducted by RealTrack Systems in the context of the FP7 EXPERIMEDIA project. The Experiment consists of the development of a Virtual Community for the CAR Venue accessible using a browser where cinematic and physiological parameters with high definition recordings (from every training session) will be available in real time and/or remotely (if needed), and with the aim to provide rapid feedback to improve the athlete's performance.

In High Performance Centers, elaborated information is highly important. Information can come from multiple devices: wearables, machines, cam-recorders or information stored in the database. There is therefore a lot of raw data that has to be processed in order to be useful. This is one of the big problems: due to the different communication protocols and because each one uses its own system.

The current state of the art brings us an opportunity to deal with the previous problems. First, because today technologies give us the chance to synchronize and deal with different devices in real time, and secondly, it gives the coach remote solutions, such as access to training sessions from his phone, no matter where he is, and in real time.

All that provides rapid feedback: elaborated and relevant information in real time and remotely if needed.

2. What is Experimedia

EXPERIMEDIA is a collaborative project aiming to accelerate research, development and exploitation of innovative **Future Media Internet** products and services through testbeds that support experimentation in the real world which explore new forms of social interaction and experience in online and real world communities.

EXPERIMEDIA develops and operates a unique facility that offers researchers what they need for Future Media Internet experimentation. EXPERIMEDIA aims to explore new forms of social interaction and rich media experiences enabled by the Future Media Internet considering the demands of both online and real-world communities associated with Live Events. This will be achieved by research, development and operation of a unique FIRE facility targeting the Future Media Internet research community working with stakeholders such as venue management, broadcasters, content providers, application developers and service providers. The approach is based on a clear definition of testbed:

"An EXPERIMEDIA testbed is a socio-technical location where individuals and communities go for experiences, learning and social interaction. It is facilitated by the Future Media Internet and must provide the four foundation elements (Smart Venue, Smart Community, Live Events and Baseline Testbed Technologies) to support experimentation into new forms of social interaction, rich media and augmented reality considering the demands of both online and real-world communities."

- Smart venues: attractive locations where people go to experience events and where experiments can be conducted using smart networks and online devices.

- Smart communities: online and real-world communities of people who are connected over the internet and who are available for participation in experiments

- Live events: exciting real-world events that provide the incentives for individuals and smart communities to visit the smart venues and to become participants in experiments.

- Baseline FMI testbed technologies: state-of-the-art Future Internet testbed infrastructure for social and networked media experiments supporting experimentation of user generated content, 3D Internet, augmented reality, integration of online communities with full experiment lifecycle management

EXPERIMEDIA has three venues for experimentation in this call:

- Schladming: an Austrian alpine resort

- CAR: a high performance athletic training facility in Barcelona

- FHW: a Greek cultural centre and museum for Hellenic culture and history

CAR Venue requirements: experiments must aim to enhance training sessions by using technology to improve the performance of athletes and help sports scientists design training plans through detailed multi-factor monitoring of biomechanics and physiology. Specific technical areas of interest are advanced techniques to manage athlete performance data acquired from cameras and multiple Wi-Fi based sensors including:

- Systems for high quality video acquisition and management targeting non-invasive real-time sports analytics and

- Multiple sensor-based performance tracking delivered via personal mobile and / or Wi-Fi sensor gateways.

- Acquired data must be synchronized and integrated to provide multi-factor views on athletes' training sessions.

As we are going to see abode, Realtrack Systems had the perfect device for these requirements and the above goals of Experimedia. And this tool is called Wimu.

3. RealTrack Systems

Realtrack Systems SL. is a Spanish technological Enterprise based in Spain focused on Sport Science, New Communications and Physical Activity Monitoring.

It was founded in 2008 and has developed wireless solutions for Auctions, Portable physical Monitoring Devices and Group workouts trainings. Realtrack Systems is the first Spanish Enterprise member of the International Alliance ANT+ (www.thisisant.com).

The Workforce is a multidisciplinary group made up of highly trained professionals. A short profile of the main individual who involved in this Experiment is above:

- ### Isabel Pérez Segura (CEO and Founder)

Linked to the University of Almería for have been studying a business bachelor, renowned for the founding in 2000 the Association of Young Entrepreneurs of Almería, in addition to representing the student community as a member of the Senate of the UAL and Central Board of the Faculty of Economics at that stage. Works in parallel in C&M Communication and Multimedia, where she gained experience in business and project development for new technologies, always relatedcomputer science and communications where she is responsible for multimedia projects, sales and the company's administration, collaborating with Carlos Padilla in virtually all her career.

- ### Carlos Padilla Sorbas (Founder, CTO)

With over 16 years experience as a freelancer in computer science, using the trade name C&M Communication & Multimedia, has done work related to new technologies for entities such as Cajamar, various municipalities, provincial governments, and other public entities primarily Multimedia, WEB, IMMERSIVE PHOTO or 360º, Proximity Marketing. 2001, awarded by the Instituto Andaluz de la Juventud on Economy and Employment. 2003, begins to develop tracking and mobility software for mobile devices, mainly for monitoring PDA mobile de- vices and Nokia (becoming one of the few Spanish members of the Nokia Forum Pro), this application is sold through Internet worldwide. He is also the Hardware Designer behind WIMU and QUIKO system.

- José Pino Ortega (Sport Area Responsible and Founder)

Doctor in Physical Education, specialized in Football and High Performance in Football. He has collaborated with various entities in the coaching of football teams. He has published 12 books and has been involved in several articles as well as in courses and conferences in Spain and internationally. He has participated in several research groups. Combines these activities with his teaching job at the University.

- Victor Andrés Andrés (Hardware and Embedded Systems Developer)

Victor is a Computer Engineer from the University of Almeria specialized in Embedded Systems and hardware devices. Victor has been part of the development team of Realtrack Mobile (a system to track cell phones) and "Realtrack Indaseñal" (a Schedule tracking system for the company INDASEÑAL). He has conducted projects focused on Linux-based Servers Management and web browsing programming. In the last years he has being working in the Design Team of WIMU. He has collaborated with University of Almería in 2009 and 2010 as lecturer for "Intercambio Directivo y Personal Program" and as Co-Project Tutor in 2013 with Doctor José Antonio Álvarez Bermejo.

- Jose Pérez Cara (Computer Programming and System Analyst)

Jose is a Computer Engineer from the University of Almeria specialized in Social Robotics. Before join Realtrack work as a Professor and contribute to the creation of the wireless network Ejido Wireless (2005). He is currently an analyst and designer of applications based on real time analyses, GIS representation and 3D modeling.

He has also being working in the programming QUIKO, a PC software focused in Sport Parameters Analysis.

- Manuel Vera López (International Manager)

Manuel is specialized in foreign trade and social marketing. He holds a degree in Business Administration from the University of Seville, a Masters in International Business Operations by the Chamber of Commerce and another master in International Trade by the EOI. He has worked as a logistics manager for the Alter Group, as a consultant for the Commercial Office of the Embassy of Spain in London, and currently, as International Manager for a Realtrack. He is the founder of foreigntrade20.com, an online encyclopedia of foreign trade with over 70 articles, guides and manuals on international procedures. He is the author of international trade and personal branding books like "How to become an honest con artist" (2014), "International Sale Price" (2014), "Selling the moto" (2014), "Step by Step Guide Incoterms 2010" (2013) or "International Payment Methods" (2013).

- Miguel Ángel López Vicente (Apps Developer)

Technical Engineer, specialist in Back-Front and Back-End Computer Programming. Miguel Ángel has programmed and launched Android Applications such as *Embolsados* (for "La Junta de Andalucía"), *Onda Campos* (for the University of Extremadura) and *Radio Universidad* (for the University of Almería). He has also developed apps and widgets for Smartphones with thousands of download.

- Raúl Gómez Salvador (Graphic Designer)

Raúl is Realtrack's graphic designer. During the project, he's been in charge of all graphic development for the CARVIREN website. He studied Arts and Design in Almería and he is specialized in Advertising Graphical Design. He's been working in Realtrack since 2013. He collaborates with other brands and companies such as Gergaleña or Enrique Gonzalo SL. He also carries out book interior and cover illustration.

4. Experiment Description

The experiment consists in the development of a Virtual Community for CAR venue where cinematic and physiological parameters with high definition recordings (from training sessions) will be available in real time and/or remotely (if needed), and with the aim to provide rapid feedback to improve the athletes' performance.

CARVIREN will use an innovative system of hubs, which collect data from Bluetooth, Ant+ and TCP/IP devices and sync them. They can be part of the venue infrastructure and when needed, become mobile transmission units, collect data, process it and send it back to the CAR venue in real time, no matter where in the world, just using Wi-Fi connection.

All these data will be accessible to different users though a virtual environment (using a browser), where different actors will have access to a series of small utilities or widgets (called Sport Widgets) that will allow them to observe the trainings and analyze the results without space or time limitation.

Sport Widgets (from now on, "Sportwis") will be varied and with different functions. Users will be able to select in between different Sportwis, depending on their roles, to customize the information they want to receive.

The whole experiment is focused on offering relevant and elaborated workout information in real time, facilitating athletes-coaches interaction and rapid feedback.

Elaborated information provides a better rapid feedback, and the possibility to perform changes at the very same time the athlete is training in the aim of a much better high performance. Coaches will not have to analyze raw data collected by different devices one by one. It would be the CARVIREN system that will process and sync all this data, which will be shown with their corresponding HD video recordings (if available).

4.1. Sport Areas of Experimentation

For this experiment, three disciplines were involved:

- Biathlon: kinematic and physiologic (Heart Rate and Saturation Oxygen in the muscle) data will be collected to check performance during the combat
- Trampolining: developing a special Application to measure the flight time and other kinematic parameters of the jump; plus physiologic analysis if wanted.
- Gymnastics: physiologic analysis using the SOm2.

4.2. Learning Objectives

Regarding the different parts involved in the experiment, there were specific learning objectives we wanted to achieve with the CARVIREN:

1. For EXPERIMEDIA:

- Test a way of promoting CAR Venue with Social Networks(using the SCC)
- Incorporated AVCC repository to our Virtual Environment.
- Use new technologies to improve performance in the CAR
- Apply new FMI tools in order to create a Virtual Network for the CAR

2. CAR Venue:

- Improve the training process, implementing remote training when needed, classes and formative material no matters where, no matters when. That is what it is called Rapid Feedback, and it's vital, for a High Performance Centre
- Develop a virtual environment that connects all actors.
- Combining devices existing in the Venue in order to have a wider vision of the workout and accomplish a smarter training.
- Combining data and video from the AV Repository and offer these information together at the same time
- Promotion using social networks and share content (approved to be shared)

3. Coaches and Athletics:

- Combine the know-how with scientific analysis and improve the workouts
- Remove physical and temporal barriers between coach and athletes.
- Establish multidisciplinary contacts

4.3. Experiment Procedure

We identified 3 phases of the experiment:

Study and Experiment set up

- Meeting with the CAR stakeholders (athletes, coaches and technical staff). Analyze and discuss the best way to proceed with the experiment. Study the documentation provided by CAR.
- Elaborate the "Experiment Problem Statement and Requirements" for the experiment.
- It is in this stage where the sample for the experiment is chosen. In this case, Trampolining, Biathlon and Gymnastic.

Execution of the Experiment

The proper development of the experiment started in November and finished in August.

Final Tests and Runs

Final test probed the feasibility of the experiment. With them, we evaluated the work done, variations and possibility for a whole scale implementation.

5. Background of the Experiment

The CARVIREN experiment focuses on high quality content production for the gymnastic training sessions. This content is what we call **relevant information**. Relevant information comes from **elaborated raw data**, including 3D motion capture based on inertial sensors and biomechanical analysis.

What is the different between raw data, elaborated information and relevant information?

The basic unit of information is the raw data. By itself, it's not useful to take decisions. It's just a series of numbers. You need to give them an interpretation.

The first step is to build that interpretation. You need to elaborate information from raw data. It can be a visual displayed or another data derivate from the raw data (speed come from space and time, for example).

Elaborate the raw data is just the first step. It's the interpretation that matters. Data and information are not important in high performance if they cannot be applied to improve the performance.

This fact requires that information need to be **relevant.**

The best way to answer it is with an example:

Raw data from an accelerometer contains metrics from each axis. These metrics can be displayed in a datasheet. Elaborated information takes that data and creates a graphical representation of the movement. Relevant information is the analysis of that elaborated information, such as number of impacts:

TIME	WIMU_163 ACELFUSION X	WIMU_163 ACELFUSION Y	WIMU_163 ACELFUSION Z
15:08:40 682	-0,0266197996156648	0,0025348525549127	-0,975304922914551
15:08:40 686	-0,0265456133392736	0,00122600677933765	-0,975168288432689
15:08:40 690	-0,0259402511468033	0,00144514074766307	-0,973300834776381
15:08:40 694	-0,024518474787791	0,00298461888869598	-0,974122184469751
15:08:40 698	-0,0247067723782096	0,00189980571098361	-0,974422513664983
15:08:40 702	-0,0239750417421187	0,00180000592436401	-0,973815171162786
15:08:40 706	-0,0237843127285997	0,00333871375975318	-0,972524509337312
15:08:40 710	-0,0238642042983846	0,00343421019267526	-0,97280129545671

Figure 1: Raw data from a 3D accelerometer.

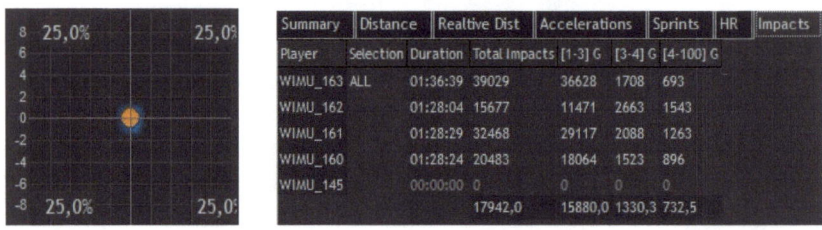

**Figure 2: Elaborated information (left) and relevant information (right).
Impacts (using a 3D accelerometer)**

For this purpose, Realtrack had the perfect tool: **Wimu.**

6. RealTrack baseline component: WIMU

The main tool for the CARVIREN experiment is WIMU, a device developed by RealTrack Systems. WIMU is a light (88x52x29mm and 120g) smart device with microprocessor, RAM memory, 3D accelerometers, 1000Hz gyroscope, magnetometer, barometer, microUSB and SD memory. WIMU has integrated Bluetooth, Ant+ and Wi-Fi radios. WIMU develops the concept of virtual sensors. Virtual sensors allow WIMU to deal with data generated by other devices like self-generated data.

Figure 3: WIMU: RTS Hub and Data Collector Device.

These features make WIMU optimal to work as a hub, by connecting multiple devices together and synchronizing them in the same network. WIMU is plug-and-play and it connects easily with the CAR Wi-Fi infrastructure. It takes only seconds to setup for the very first time. Microprocessor and internal memory allow an installation of CARVIREN inside the WIMU. This possibility is a key point for Mobile Transmissions Units.

WIMU is not only a hub, it can be more. It can be a whole **Mobile Transmission** Unit (MTU). WIMU's microprocessor can process data without being connected to the server. Internal memory and external memory (up to 32GB) can save the information collected and processed.

CARVIREN carried out research on synchronization of motion capture data gathered from the inertial sensors, video obtained from the cameras available at the CAR Venue and metadata. The experiment focused on the use of all those elements in the training process and on the improvement of the athlete's technique.

The experiment took place in the CAR Venue, with the interaction of athletes, trainers and other professionals involved in preparation and performance analysis.

7. EXPERIMEDIA Baseline Components

CARVIREN experiment used three modules developed within the EXPERIMEDIA facility:

7.1.Experiment Content Component (ECC)

CARVIREN has integrated the ECC component in order to measure data that helps us analyze the CARVIREN usage and the success of the experiment. The ECC is provided with APIs that can be used by ECC clients in order to communicate about experiments and measurements while the data is exchanged via a RabbitMQ message bus. These client APIs are available for Java, Android, C# and C++ clients. We have used the Java client to connect Baseline Components plus our Hubs.

The first task to integrate the ECC was installing it a local server at RTS installations. First tests were done locally, in a controlled environment. After installing all ECC packages in a local Maven repository, we began the integration of the module with the

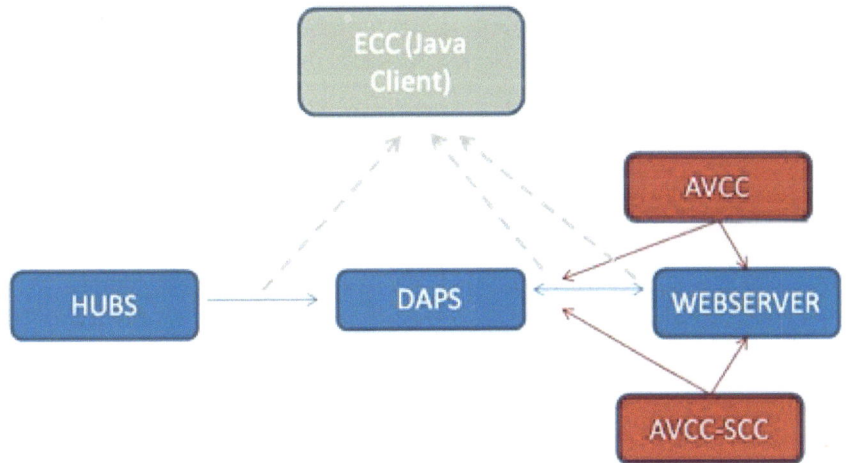

Figure 4: Diagram of the ECC and the relationship with the rest of components.

ECC was in the background, helping us to collect metrics from the Baseline Components, the website and the hubs. These hubs were not directly connected to the ECC, but to the DAPS. ECC gave us data of interest like connections, average time in the website; number of users created and connected, number of hubs connections, load workouts, created videos (and bitrate) and social metrics.

After the experiments concluded, the measurement data collected by the ECC can be accessed all at once via two CSV files (one with the metadata and the other containing the actual measurement data).

7.2.Audio Visual Content Component (AVCC)

CARVIREN accessed recorded content stored in the venue and live streaming using AVCC module provided by EXPERIMEDIA. Video can be recorded and stored at the "AV repository" of the AVCC.

The AVCC component is accessible as a webserver and offers the following interface:

Table 1: AVCC table of parameters

1. *Parameter*	2. *Description*	3. *Occurrence*
4. Title	5. A string representing the title of the video file to share on Facebook.	6. M
7. description	8. A string representing the description of the video file to share on Facebook.	9. M
10. Category	11. A string representing the category of the video file to share on Facebook.	12. O
13. multiQuality	14. A Boolean representing whether the video file has transcoding profile 'flash'	15. C
16. html5Mp4U	17. A Boolean representing whether the video file has transcoding profile 'html5-mp4'	18. C
19. html5WebM	20. A Boolean representing whether the video file has transcoding profile 'html5-webm'	21. C
22. html5OggM	23. A Boolean representing whether the video file has	24. C

1. Parameter	2. Description	3. Occurrence
	transcoding profile 'html5-ogg'	
25. html5Mp3audio	26. A Boolean representing whether the video file has transcoding profile 'Mp3 audio'	27. C
28. userName	29. A string representing the title of the video file to share on Facebook.	30. M
31. Path	32. A string representing the url of the video file to share on Facebook.	33. O

"M" (mandatory parameter) "O" (optional parameter) "C" (conditional parameter)

For the CARVIREN experiment, we have used the mandatory parameters, the transcoding profile 'html5Mp4U' (for computers) and the html5WebM (smartphones).

In order to sync data from the workout with the videos we had to request a timestamp from the video. We measured duration of the video and the workout. Because the order to start and stop both of them was given at the same time, durations of video and workout were the same. We just applied an offset[1] to both timestamps.

---

[1] In computer science, an offset within an array or other data structure object is an integer indicating the distance (displacement) from the beginning of the object up until a given element or point, presumably within the same object. The concept of a distance is valid only if all elements of the object are the same size

7.3.Social Annotation Service

CARVIREN could also be defining a professional-social network in a small scale. The aim of using this component is to test if it's possible, for a High Performance Centre, the promotion through the social networks; or at least, the use of this environment as a social network inside the venue.

Because of the nature of the installation, the confidentiality of the training sessions can be crucial. However, coaches and technical staff could identify and share non critical content (with authorization of the athletes). In addition, we have the opportunity to test interactions between all actors involved in the experiment (athletics, coaches, staff, etc.) and see how it works **in a small scale.**

Figure 5: Facebook Event Page

Here is where the integration of the SCC and AVCC modules comes in. This is how it works: a video is recorded inside the CARVIREN website (a training session or just a video); AVCC processes the video, and takes a highlight picture representative of the video. Then, a link is sent to the Facebook event page, and posted to the people invited to the event. The link cannot be shared, it doesn't appear in the user's biography, and the video still remains inside the CAR Database.

This is a diagram of the process:

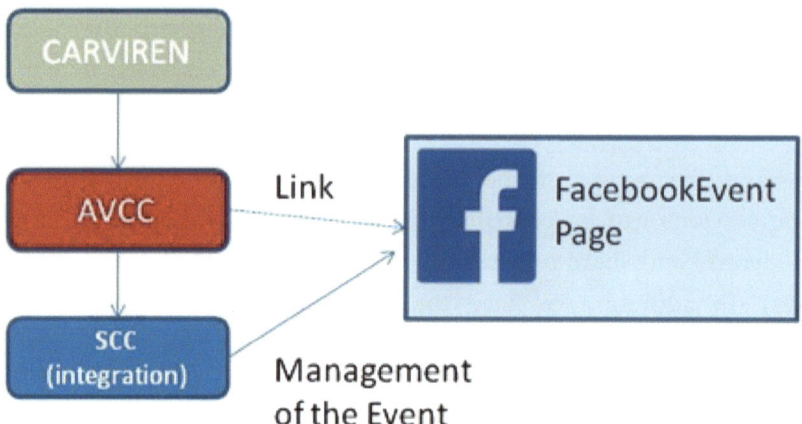

Figure 6: Integration of AVCC/SCC with CARVIREN.

Every time a new video is available for sharing on Facebook, a process running in the CARVIREN webserver sends an instruction to the AVCC webserver endpoint with both of the above mentioned parameters. Upon receipt of instruction, the AVCC webserver responds with a message regarding the status of the request.

Part 2:

Components developed by RealTrack Systems

8. REALTRACK Developed Components

8.1. WIMU: Mobile Transmission Unit

Access to the dashboard was not restricted to be in the range of CAR's Wi-Fi network. It was possible to upload a training session from outside the Venue. WIMU has been used as a Mobile Transmission Unit. The device stored all data collected from internal and external sensors onto a microSD card.

We had two scenarios:

Scenario 1: there was a network (with internet) and the hub is connected to it.

WIMU sent all the data to the acquisition and process server of CARVIREN. The training was displayed in real-time in the CARIVREN's website.

Scenario 2: no network and no internet.

In this case, a "log file" was created and stored into the SD card. In this case, to record the session, the user had to push the record button of the WIMU manually (by pushing once at the beginning and twice at the end). When the user gets a computer with internet, he just had to takeoff the SD card, insert it into the computer and load the log using the "MTU Upload Loggers Application" from the CARVIREN website:

Figure 8: MTU Upload Loggers Application

This remote functionality was tested during the second run and for the final demonstration.

We conduct a depth review of this feature in Chapter X.

8.2.Sport Widgets (Sportwis)

So far, we knew the type of information we wanted to show (relevant) and the tool we were going to use to collect raw data. What we needed was to develop a way to display this relevant information in a simple way.

The answer to this requirement was to create the Sport Widgets (Sportwis). Sportwis are small utilities that show relevant information in a simple and easy way. Sportwis run in a web browser allowing any authorized person to access information from his tablet, laptop or a Smartphone (regardless of the model or operating system).

Sportwis were available in two modes: **real time** and **delayed mode**. In real time mode, information was displayed a few seconds after it was generated, while in History mode any information could be displayed by the system according to previously established access criteria (coaches, athletes, federations, etc.).

8.2.1. Sport Widgets List

We developed other widgets according to the different needs of the experiment and the information we wanted to display for each discipline:

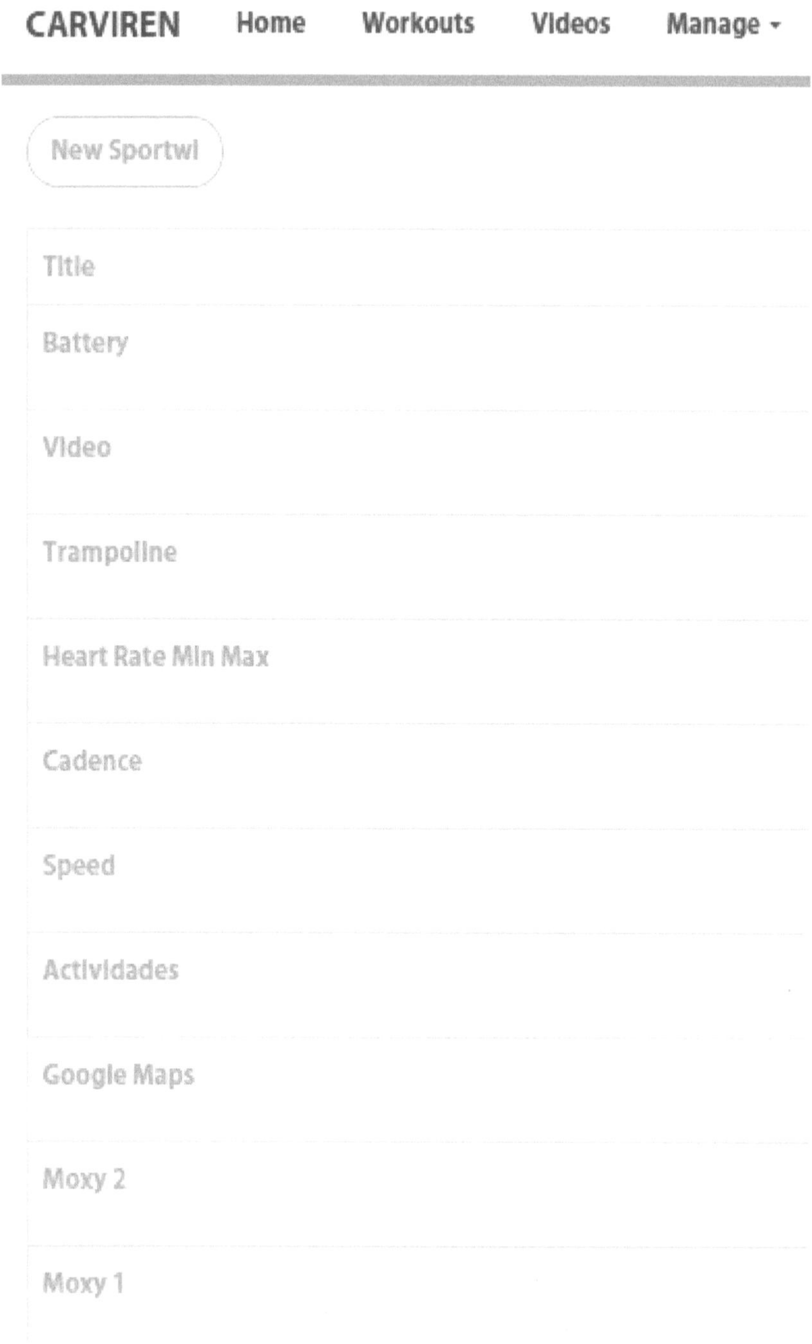

CARVIREN   Home   Workouts   Videos   Manage ▾

(New Sportwl)

Title

Battery

Video

Trampoline

Heart Rate Min Max

Cadence

Speed

Actividades

Google Maps

Moxy 2

Moxy 1

Figure 7: Sportwis list

8.3.CARVIREN website

The CARVIREN website was developed using a JSF, a framework in Java language focused on web development. The reason was simple, this language allows a better access to our DAPS and also reuses components based on HTML5. JSF provides compatibility with smartphones, tablets and computers.

The website integrated WIMU's hub system and Experiment Baseline Components (Experiment Content Component, Audio Visual Content Component and the integration of the AVCC with the Social Content Component).

CARVIREN website had also a role system where all information (elaborated information) was available in real-time and in delay mode for the different actors. This information was displayed using sport widgets and it was different depending on the user and his role.

8.4.Experiment Areas

For the experiments, four areas have been discussed with the CAR venue:

Trampolining area; where athletes had been training and recording the sessions with ATOS Camera:

Figure 8: Trampolining

Biathlon: with the study of oxygen saturation in the muscle (%), cadence, speed and heart rate parameters during cycling.

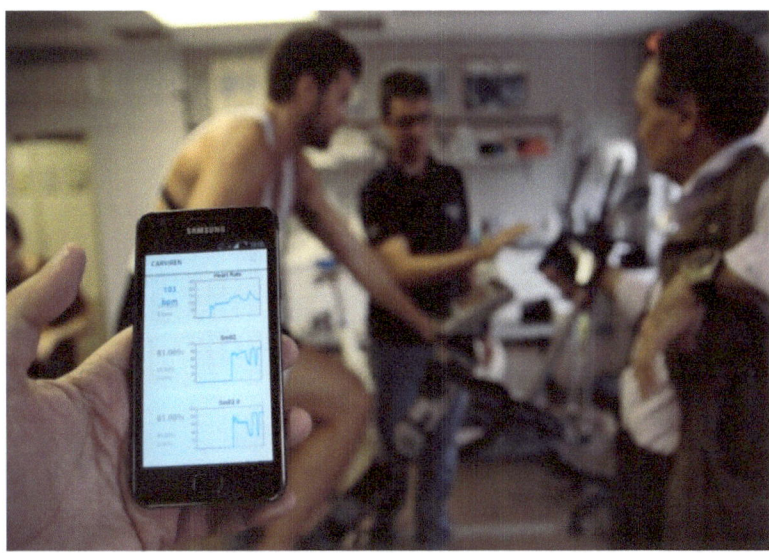

Figure 9: Moxy studio

Gymnast: occasional tests in different areas, such as pommel horse with Moxy (the device that measures oxygen saturation in the muscle Sm02[2]).

Figure 10: Pommel Horse test

---

[2]SmO2 tells you how much oxygen is left after the tissues have taken the oxygen they need

Remote Training: when an athlete is outside the CAR, we can still record session with WIMU and external sensors. They are saved inside the SD card of the device and loaded to the CARVIREN website:

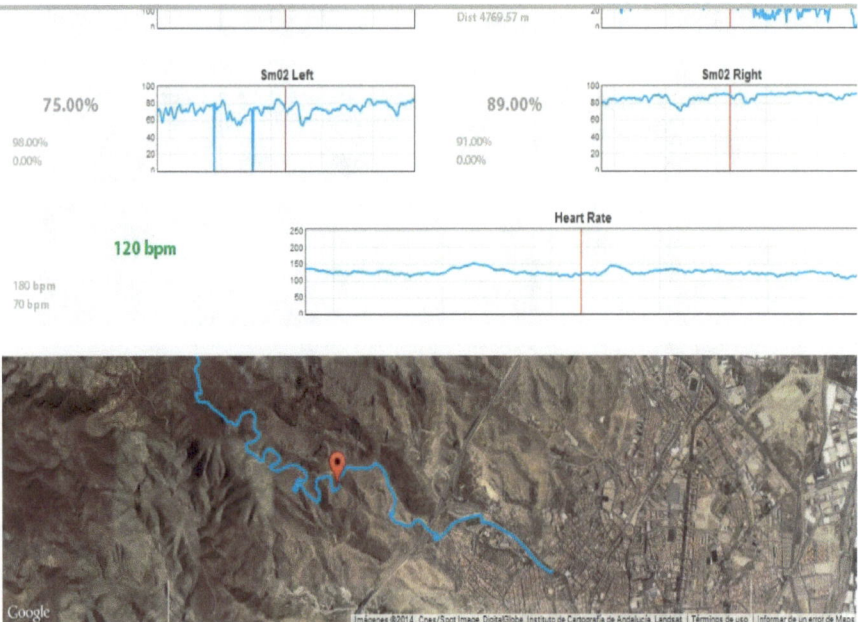

Figure 11: Remote uploaded workout in bicycle.

9. Depth Review of the Trampolining Widget

Trampolining widget is a special widget developed for this experiment. This widget measures the flight time and can identify a jump. Why is this parameter so important?

Time-of-flight machines measure how long trampolinists spend in the air, and that calculation is factored into the scoring. This widget uses inertial sensors from WIMU (3D accelerometers, Magnetometer and gyroscope). Using a selection of each sensor's channels, and applying a logarithm; an accurate flight time can be calculated.

It also measures maximum jump acceleration in Gs and angular speeds for the three axes.

Validation was made with HD videos (in slow motion) and the *Trampoline Synchron Device* (used in competitions).

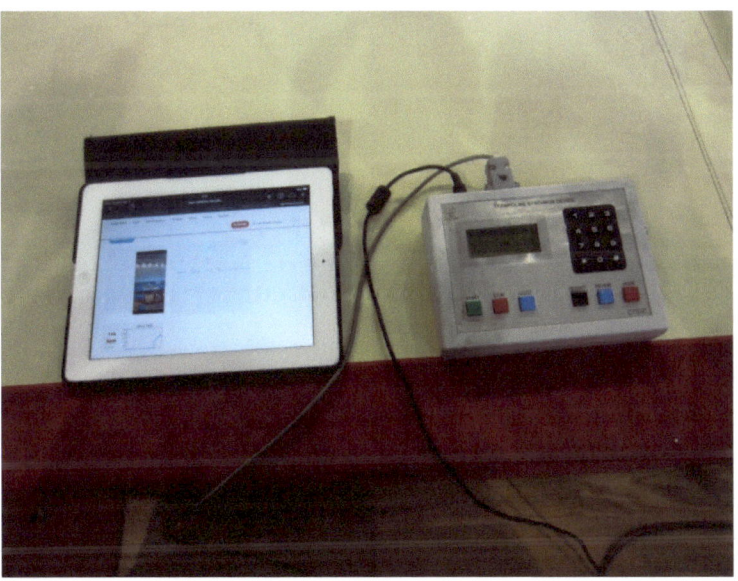

Figure12: Analysis of the 4th Jump of an athlete during a session (video and trampolining widget)

#	FT ms	MaxG g	X °/s	Y °/s	Z °/s
1	1200	4.29	0	0	0
2	1290	5.28	0	0	0
3	1310	5.51	0	0	0
4	1350	6.78	335.88	263.01	227.83
5	1340	6.84	0	290.14	0
6	1360	7.79	308.13	306.06	189.66
7	1460	7.52	0	429.96	119.17

Figure 13: Analysis of the 4th Jump of an athlete during a session (video and trampolining widget)

This was the final result. Behind this widget there was a meticulous research we are going to review now:

The purpose of this study was to determine the flight time variable on trampolining. This variable is related to the *takeoff acceleration, angular speed* during the jump and *landing impact*.

Following we can see the different phases for a trampolining jump:

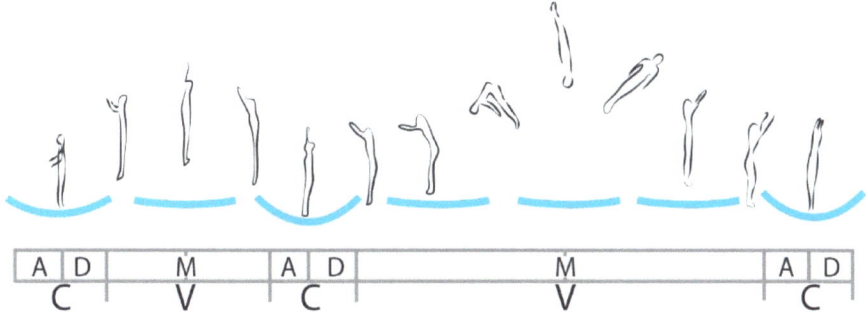

Figure 14: Analysis of the 4th Jump of an athlete during a session (video and trampolining widget)

Where:

*C is the contact phase divided in two sub phases: **A** (landing) and **D** (takeoff).*
*V is the flight time phase. Maximum peak (where the athlete reaches the maximum height) is **M**, and it splits V into two parts.*

The **first step** was recording a session in realtime using WIMU. Our references to test accuracy were two: video from a HD slow motion camera and the official device used in competitions.

After several tests, we found out the best place to put the Wimu was in the lumbar region (close to the mass centre):

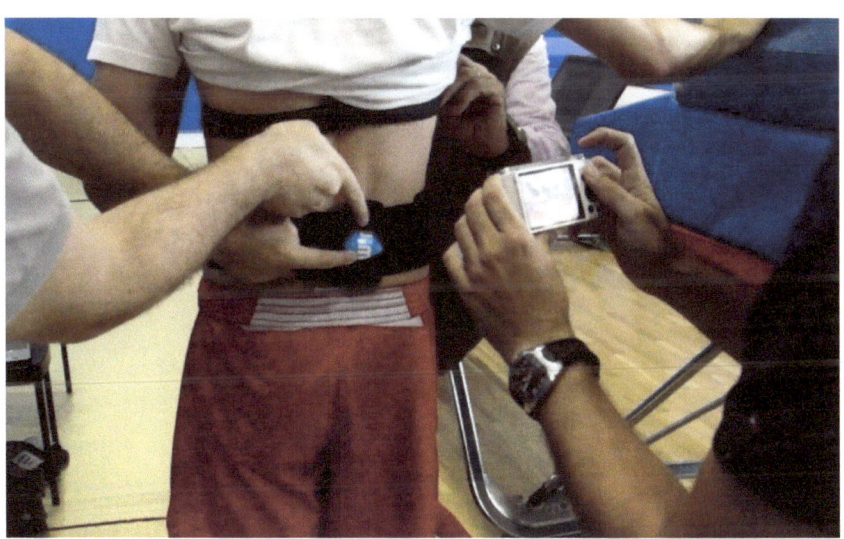

Figure 15: Analysis of the 4ᵗʰ Jump of an athlete during a session (video and trampolining widget)

The belt was specially designed for this occasion.

After recording the session, we could establish a pattern using the Attitude sensor of Wimu (that integrates different channels from accelerometers, gyroscope and magnetometer):

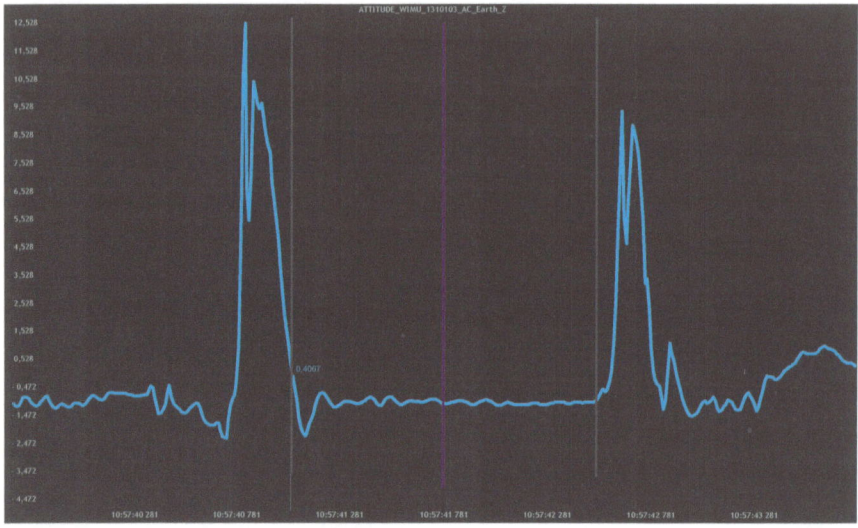

Figure 16: Analysis of the 4ᵗʰ Jump of an athlete during a session (video and trampolining widget)

This was the chart of an athlete's jump registered with Wimu. Blue line is the jump phase (acceleration). Horizontal red line is for the takeoff phase. Green line coincided with maximum peak where maximum height is reached. And finally, yellow line is when landing phase start.

After analyzing and defining the pattern, the next step was automatization. The system needed to find and separate all jumps inside the session. We developed an automatic logic logarithm for segmentation. Automatic segmentation is shown in the following image:

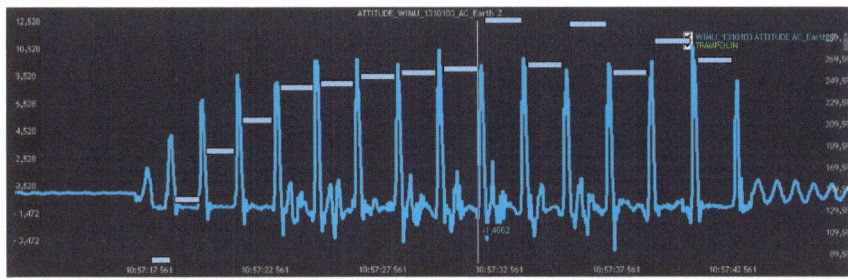

Figure 17: Analysis of the 4th Jump of an athlete during a session (video and trampolining widget)

In the chart, jumps are displayed with horizontal blue lines that cover total flight time for each jump during the whole trampolining routine. High values of acceleration happen during the takeoff and landing phases. With a minimum and maximum time, and acceleration tolerances conditions; we can classify the jump correctly.

Result of the logarithm can be verified watching the footage from the synchronized video. We can take frames that correspond to each phases defined for the logarithm (takeoff, maximum height and landing). If we analyze these moments with the logarithm, we can see how they coincide with the frames:

10.   Depth Review of the Moxy Widget

Moxy monitor is a device that measures Oxygen saturation on the muscle. Moxy uses light from the near-infrared wavelength spectrum (light from about 670 to 810 nm) to measure muscle oxygenation levels in muscle tissue. Human tissue has low optical absorbance of near-infrared light, so the light can travel to reasonable depths. The near-infrared wavelength range is particularly useful because hemoglobin and myoglobin change colour in that range depending on whether or not they are carrying oxygen.

SmO2 stands for muscle oxygen saturation. It is the percentage of hemoglobin that is carrying oxygen in muscle tissue. Hemoglobin is the molecule in red blood cells that actually carries oxygen from the lungs to where it is needed in the body. Hemoglobin responds to chemical signals to drop off oxygen where it is needed. Hemoglobin's most common states are oxy and de-oxy. The measurement of SmO2 takes place in the capillaries of the muscle. This is where the oxygen is being consumed. SmO2 can be thought of as a measure of the balance between supply and demand for oxygen in the muscle. When you first start to exercise, oxygen demand increases, but the heart hasn't had a chance to speed up, and the blood vessels in the muscle haven't dilated. The SmO2 drops quickly in these conditions.

As you warm up, your heart rate increases, and the blood vessels in the muscle dilate to 2 levels. When you stop exercising, the demand for oxygen suddenly falls, but the heart rate is still elevated and blood vessels dilated. At this time, a rapid increase in SmO2 is observed.

Generally, higher levels of exertion in the muscle lead to lower SmO2. The SmO2 value is also affected by other factors such as the hemoglobin dissociation curve shifting and other chemical and neurological factors.

For the incorporation of Moxy to the experiment, the first step was to integrate this device in our main core product: Wimu, so we could show it in Quiko:

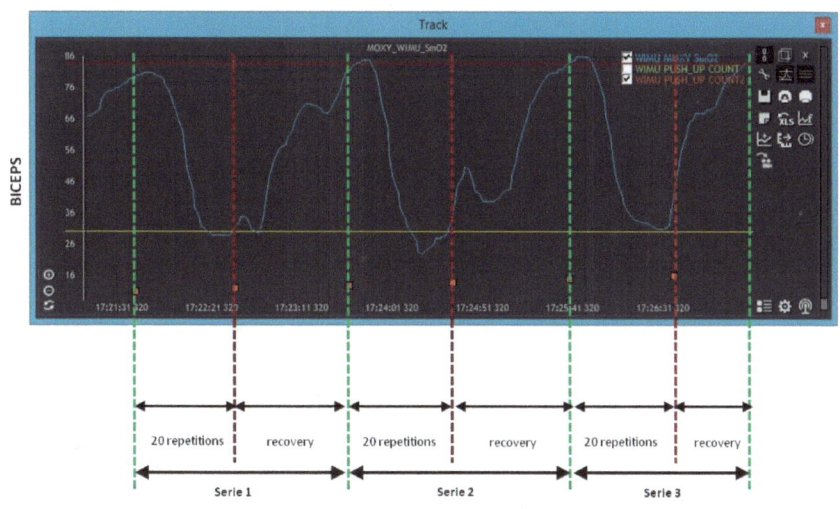

Figure 19: Visualization of Moxy in Quiko using WIMU

Second part of the incorporation of Moxy to the CARVIREN was to create the Widget for a web-based system.

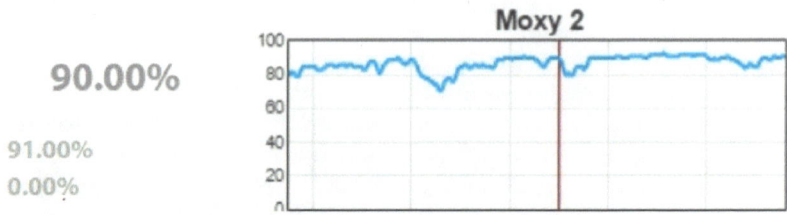

90.00%

91.00%
0.00%

Figure 20: Moxy Monitor

If we check Moxy monitor for training when the routine is being repeated, we can see how the oxygen saturation is decreasing during the exercise and how much time it takes to recover former state. With this purposes, we created a mark system, so the coach could set the beginning and the end of a routine:

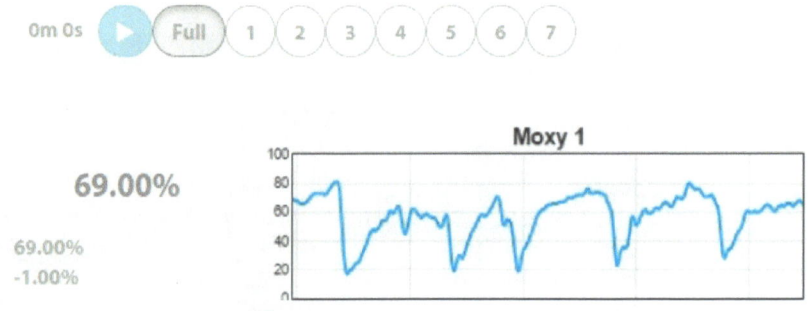

69.00%

69.00%
-1.00%

Figure 21: Full session with moxy

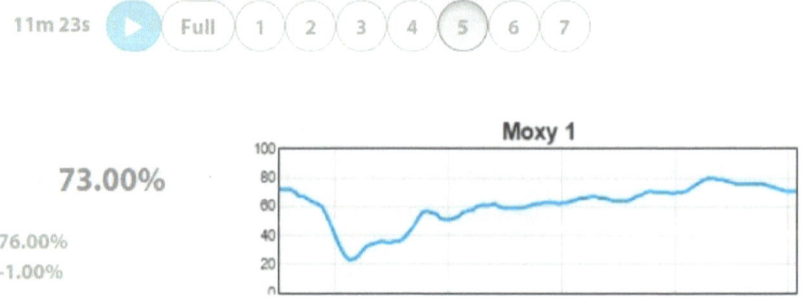

73.00%

76.00%
-1.00%

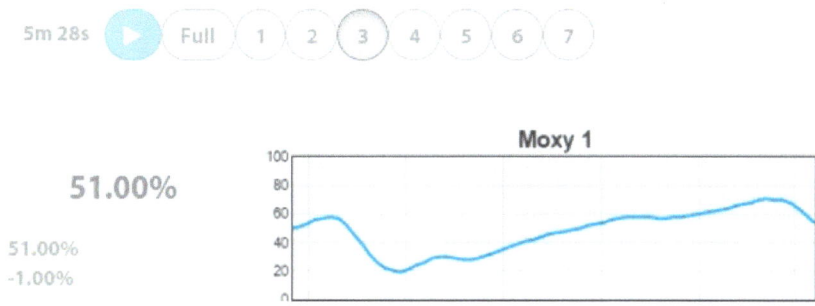

Figure 22: Number of repetition (of the same routine) in the session

Interpretation was as follow:

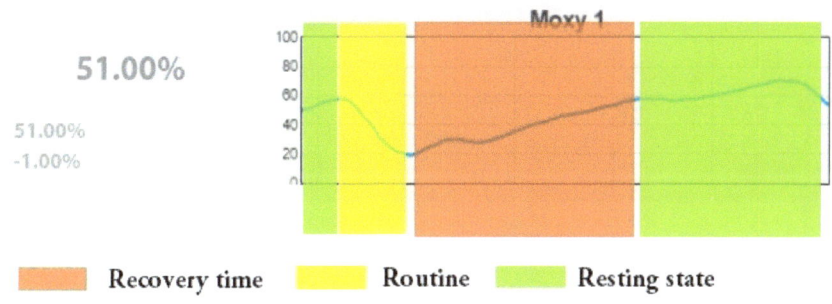

Figure 23: Routine interpretation

Using the marks, the video and the moxy monitor, we could analyze each repetition:

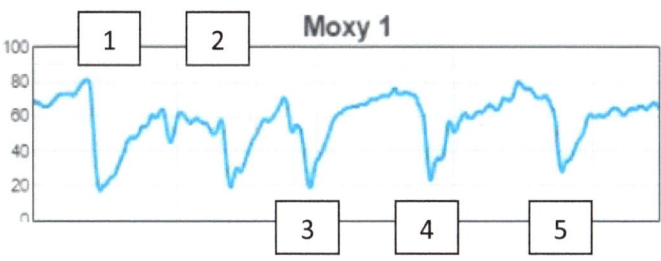

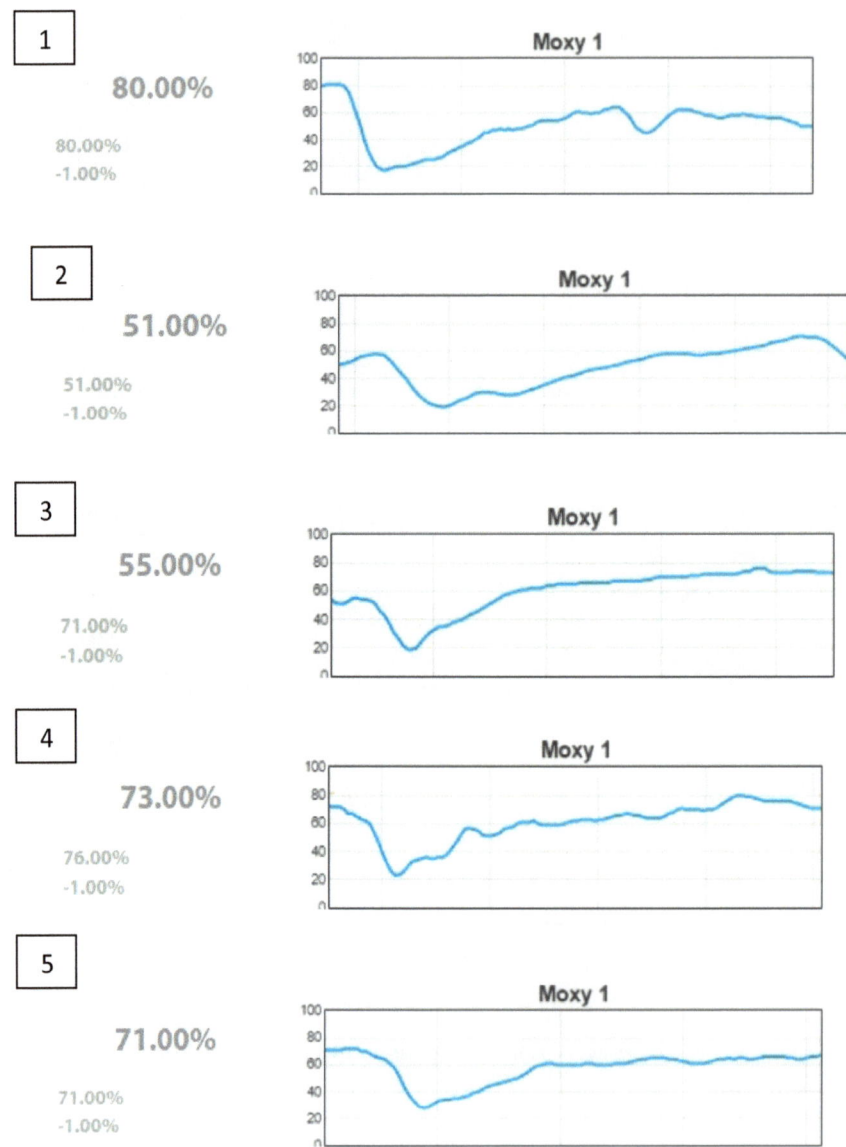

Figure 24: Routine repetitions

Part 3:

Experiment Execution

11.   Experiment Architecture

For this experiment, we have used three Baseline Components plus experiment-specific components developed by RTS. We are going to explain briefly each one of them (a more detailed description can be found in Deliverable 4.12.1) and we are going to review the architecture of the experiment and the components.

The diagram in Figure 22 gives a general idea of all the components involved and its relations:

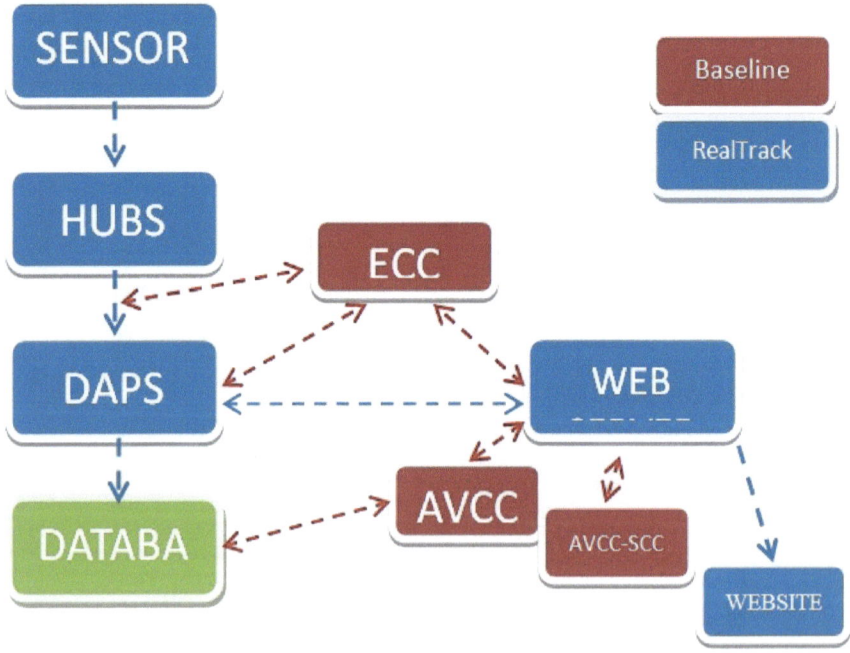

Figure 25: Architecture for the CARVIREN experiment within EXPERIMEDIA.

11.1.   RealTrack Components

In order to create this virtual environment, the following components have been developed:

11.1.1. Sensors

The first step was to establish which devices we are going to use and which ones needed a specific implementation study of their API/SDK. For example, we integrated Moxy, the oxygen saturation in the muscle monitor, speed and cadence sensors. These three sensors, they all used ANT+ communication protocol.

The new sensors required modification of the firmware of WIMU. WIMU could support 2 external profiles (two heart rate belts). With the new firmware, it was updated up to 8 external sensors, which is the maximum supported for ANT+ protocols (cadence, speed, heart rate, Moxy and remote control).

11.1.2. **DAPS**

The Data Acquisition and Process Server (DAPS) is the system in charge of reception and processing of data in order to generate elaborated information. This information will be stored in the database and it is available to the CARVIREN Sportwis.

DAPS has been installed in the CAR data centre. We have a space already configured with 100Gbs, 5Gbs RAM, Quad core and the Ubuntu 14.04 LTS I.O. installed.

DAPS manages the database with the sessions. Information arrived to DAPS by the information collected by the sensors linked to WIMU (internal and external). DAPS creates a session and saves it in the database, so it's able to provide "sessions on demand" to the CARVIREN users. Users connect to the website. The website connects to DAPS using web sockets.

This is the UML Class Diagram:

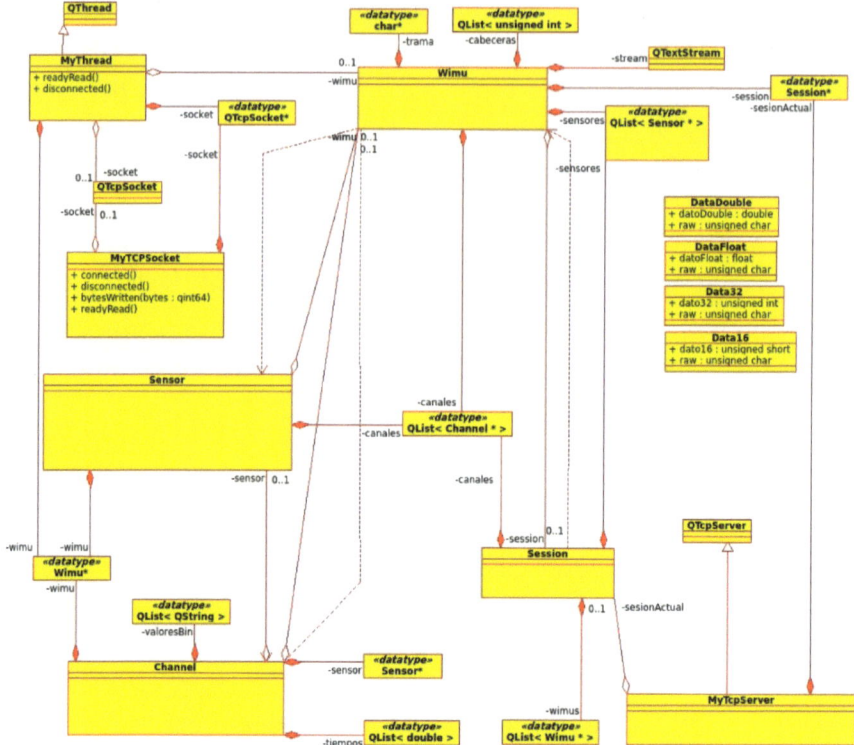

Figure 26: UML Class Diagram of the DAPS

11.1.3. Sport Widgets

Sportwis are developed in Dart. Dart is an open-source web programming language. Dart compiles the source code into JavaScript code, which allows it to be compatible with all modern browsers.

Communication between the CARVIREN website and the Sportwis is done by web sockets, allowing a bidirectional low latency communication. Once this phase of the experiment was completed, we integrated the Sportwis with the rest of the system.

11.1.4. CARVIREN Website

The CARVIREN website was developed using a JSF, a framework in Java language focused on web development. The reason was simple, this language allows a better access to our DAPS and also reuses components based on HTML5. JSF provides compatibility with smartphones, tablets and computers.

11.1.5. RealTrack Systems Component Architecture

Following, there is a diagram of the architecture of the CARIVEN experiment. This diagram shows how the RealTrack components were installed at the CAR Venue and how they communicate:

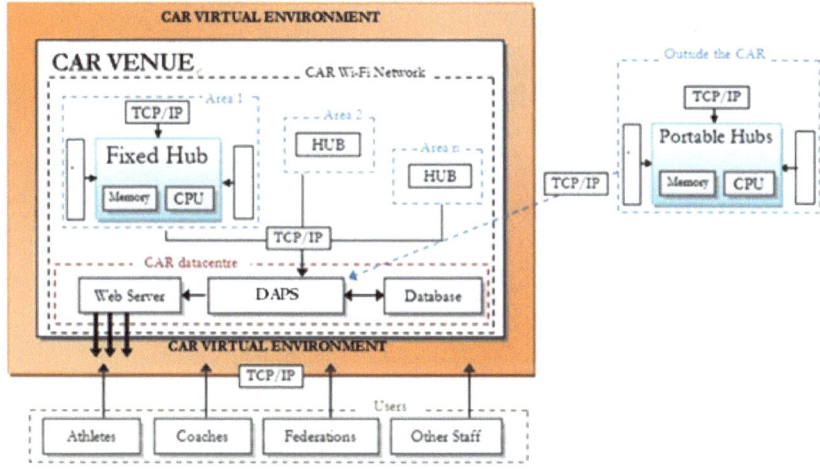

Figure 27: Low-level system architecture for the CARVIREN experiments within EXPERIMEDIA.

12. RealTrack Components and EXPERIMEDIA Components

Following, there is a diagram of the relationship between RealTrack components and EXPERIMEDIA Baseline Components:

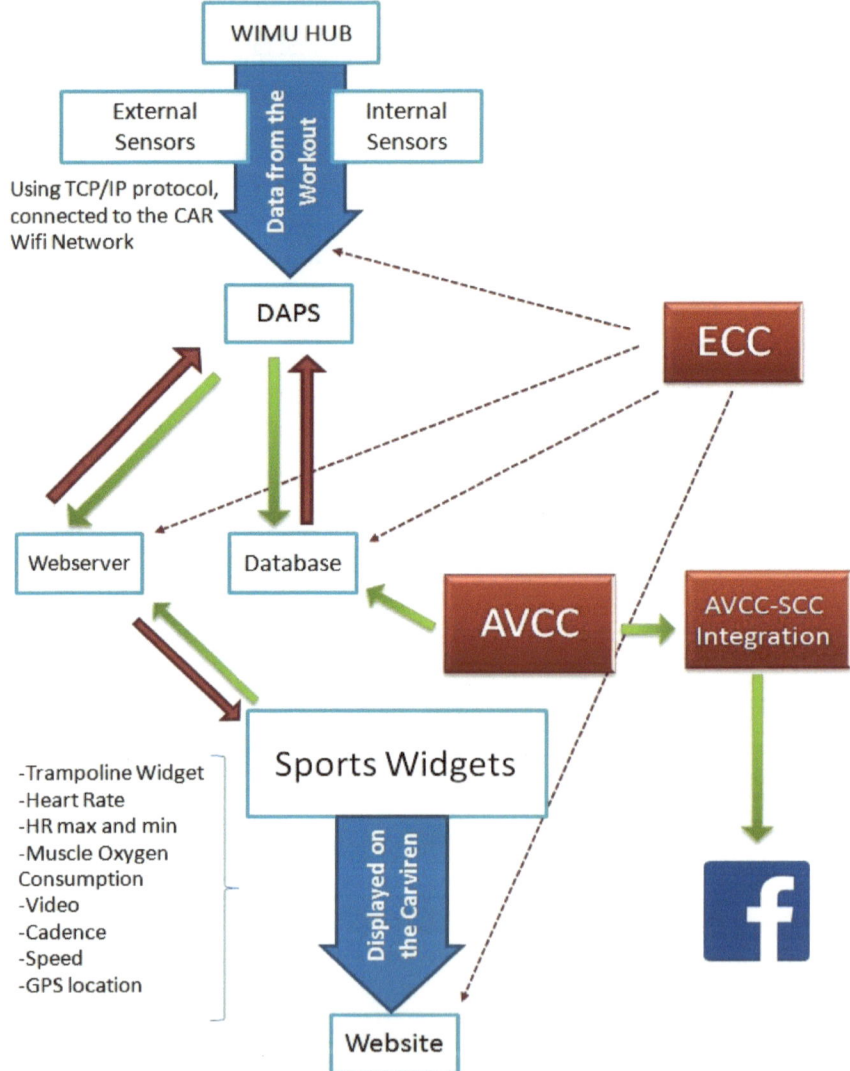

Figure 28: Relationships between RTS and EXPERIMEDIA components.

12.1. Functionality of the System

The integration of RTS and EXPERIMEDIA components gave us the following functionalities of the CARVIREN:

- Live and delayed recording of trainings
- Users role manager
- Sportwis: Moxy, heart rate (max, min and average), trampolining flight time, maximum jump acceleration in Gs and angular speeds for the three axes, cadence, speed and track in Google Maps.
- Video recording with AVCC camera
- Notifications on Facebook Event Page of each recorded AVCC video.
- Remote training loading
- Result exportation (in xls)

13.   Experiment Preparation

This section outlines the preparation necessary to perform the experiments.

13.1.   Participants and Venue

The experiment has taken place in the CAR venue. Over a period of 2 months, we have performed two experiment rounds and two free use weeks.

The **first run** took place the last week of July (starting the 21st of July).

The experiments involved the following participants:

- Coaches from trampolining who used CARVIREN during the first run (2 people).
- Athletes (5 people): three from Trampolining, one from Triathlon and one from Gymnastic. CAR technical Staff and Scientifics (7 people).
- RealTrack Systems Staff: experimenters have overviewed the execution and analysis of the experiments.

The **second run** took place the first and the second week of August (from 4th to 14th):
- One coach from pommel horse discipline.
- One athlete from pommel horse too.
- CAR technical Staff and Scientifics (7 people).
- RealTrack Systems Staff: experimenters have overviewed the execution and analysis remotely from Almeria.

13.2.   First Run

We were testing functionality for trampolining and oxygen in muscle saturation. We used the ECC to collect QoS data from the baseline components and CARVIREN data of interests. For the experiment,

CARVIREN data of interest were time spent in the website; number of users connected and number of training recorded.

System set-up involved various tasks:

1. Connection of WIMU to the Network.
2. Set up all the components for the workout (HR Belt/Moxy; fabric belt, instructions, etc).
3. Creation of a new training in the CARVIREN system (Figure 13)
4. Start the workout
5. Run the workout (Figure 14)
6. Stop the training
7. Review of the training in delay mode (Figure 15)
8. Verbal feedback from athletes and technical staff.

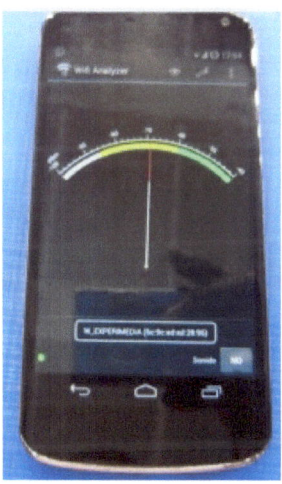

Figure 29: Scanning quality of the signal.

CARVIREN   Home   Workouts   Videos   Manage ▾   Guide

Name

Training 8/14/14 11:47 AM

Discipline

Trampoline   ▼

Athlete

---   ▼

| ---- |
| **Marc Torras** |
| Claudia Prat |

---   ▼

Moxy 1 in

Moxy 2 in

☐ Camera

Figure 30: New training form

Figure 31: Josep Escoda with Clara Bozzo (Trampolining Coach)

Figure 32: Post training analysis in the CARVIREN Website

13.3. Second Run

The second run was conducted during the second week of August. It helped us to solve some problems with the ANT+ connection. We solved them by updating the firmware of the WIMU. For this run, there were two tests scheduled: a pommel horse session and the remote functions test.

We started with the pommel horse session. For this test we measured oxygen saturation in the muscle and recorded a workout.

Figure 33: Pommel horse test

This time, and thanks to the first run, we didn't need previous steps 1 and 2 from the first run

The rest of steps were the same:

Set Moxy for the workout
Creation of a new training in the CARVIREN system
Start the training
Run the workout
Stop the training
Review of the training in delay mode
Verbal feedback

After the pommel horse session, we continued with the second part of this run: the remote functionality of CARVIREN. In this part of the experiment, WIMU acts as a Mobile Transmission Unit

We tested the two scenarios mentioned in section2.1.3. For both tests, we used the help of Victor Andrés (worker of RTS and member of the Sport Association Mastrinkais).

First Scenario: WIMU outside the CAR, connected to a wifi network with internet access. Podcasting in real-time from Almería:

The steps were:

1. Simulate that the RTS's office was a sport venue
2. Connect WIMU to our wifi network
3. Add the external sensors (heart rate, cadence, GPS, speed and oxygen saturation %)
4. Log into the CARVIREN's website
5. Select the remote WIMU (Almería)
6. Start a workout (for triathlon)
7. Go around the block with the WIMU
8. Stop the workout
9. Review

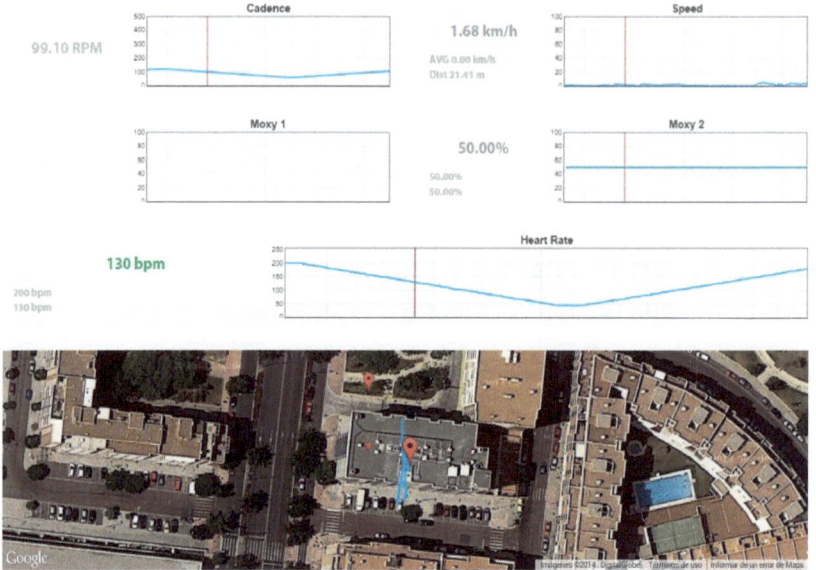

Figure 34: MTU test. Scenario 1 (outside the CAR connected to a network with internet access)

Second Scenario: WIMU outside the CAR with no network or internet connection.

We used two Moxys, cadence, speed sensor, heart rate belt and gps.

The steps were the following:

1. Fully charge the sensors and the WIMU
2. Check SD card inside the WIMU
3. Preparation of the sensors on the bike
4. Set the Moxys
5. Go to an outdoor location (Sierra Alhamilla, Almería)
6. Start the session manually (pushing once on WIMU's record button)
7. Ride and carry on with the workout
8. Stop the session manually (pushing twice on WIMU's record button)
9. Save the log file from the SD card into the laptop

10. Log into the CARVIREN website (with admin role)
11. Go to the "upload logs tool", and load the workout
12. Review of the workout

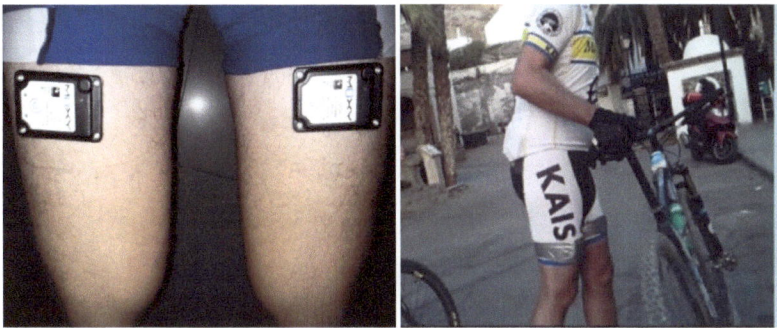

Figure 35: Preparation for the remote test

Figure 36: Review of Victor's workout

The results and analysis of the two runs can be found in the next section.

14. Experiment Execution

Each experiment run was accompanied by its own steps guide (seen in the previous section), which aided the experimenters to ensure an efficient and correct execution of the experiments. Relevant details of the execution for both experiments are reported in the following paragraphs.

14.1. First Experiment Run

We had a problem connecting the web-service to the ECC, so we could not take any data about average of usage of the website. We did take data for the rest of components.

We also experienced some issues regarding to the roaming and the hubs. RTS had to update some parameters from the WIMU and update the firmware. Once we solved this problem, we proceeded with the first part of the run.

The coaches and the athletes in trampolining conducted a series of workouts, in particular series of 10 jumps. We recorded all these sessions, trying to have a sample for each different jump (for later adjustments). Data was not fully synchronized with the video recorded by ATOS camera.

Two future improvements would come from this first part of the run (the timestamp from the video and an accurate flight time thanks to the sample).

The second part of the run included a Moxy in each leg, during the cycling workout of a triathlon test. Several CAR doctors and an athlete took part in this test. We recorded a session with two Moxys, the heart rate and the cadence sensor. The session was replayed after being recorded.

Feedback from the doctor at the test was good. Sm02 was definitely a very interesting parameter for them. So far, there are no other devices that measure this oxygen saturation in the muscle. This oxygen consumption is a key-parameter for improvement. Two Moxys allow a differential analysis for two opposite parts of the body (right and left leg; right and left arm, etc). The measures of Moxy (smaller than a Smartphone), the results in real-time and the export tool to xls were the perfect widgets for this session.

Figure 37: Moxy 1: left leg; Moxy 2: right leg

14.2.   Second Experiment Run

In this run the ECC connection was available and we could monitor in real-time data about average use. The run has two parts.

For the first part (the pommel horse test), we recorded a training with a duration of 20 minutes. The workout consisted in the repetition of a pommel horse routine. For this test, we add a new feature, the **activity marks**.

This new feature allowed the coach to make marks between routines repetition:

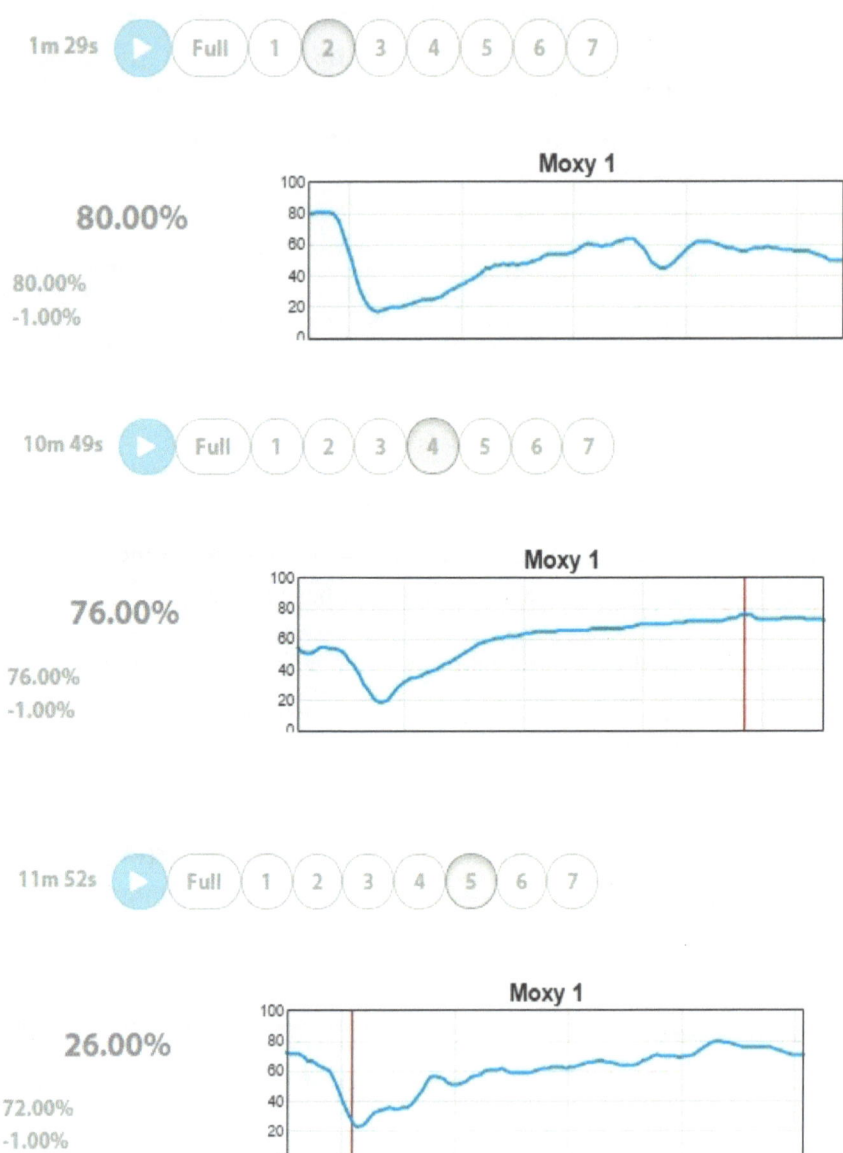

Figure 38: Activity marks during pommel horse test of run 2

No problems were detected during the test. Coaches keep analyzing the session after the athlete finished for at least 60 minutes. Once again, the oxygen saturation measurement in the muscle widget was one of the most appreciated features.

The second part of the Run (the MTU scenarios) was conducted without issues. We uploaded four sessions with different sizes: 5mb; 15mb; 50mb and 100mb. A 100mb session contains almost 70 minutes of training with a frequency of 40Hz

15.   *Quality of the Experiment Conclusions*

Overall, the main activities and functions of the CARVIREN worked.

We experienced some issues. A close review of each component follows:

- **CARVIREN website**: the access to the website was good. There were no problems with speed. All but one of the users responded fast or very fast (87.50%). One of the biggest problems of websites is the access time. Being fast doesn't guarantee success, but a slow access is unforgettable.
- **AVCC Component**. More than the 80% of users qualified it as good/very good (7 or more in a scale of 1-10). The visualization was good too (71.4%). Thanks to the survey, we realized it was a good practice to reboot the server (it takes less than 2 minutes) before starting recording sessions (once a week top) to ensure it functions properly.
- **Experience on the** Social Annotation Service is also a 100% positive. The sample in Facebook included the 3 athletes that participated in the survey plus 2 of the other 5 users. All of them thought the service was good and fast enough. Again, with this kind of technologies, we are dealing with access time (or in this case, display time), which needs to be fast so as not to diminish the QoE.
- **User friendliness of the CARVIREN.** Thanks to informal feedback and after using the website, we realized there were some issues with user friendliness. We did some changes to improve the usability.
- There were two kind of access to the CARVIREN. For technical one (for developers) it was very tricky to make changes and, because we were dealing with functions such as restarting the server, connecting to the DAPS, or monitoring in real-time the web environment of each Hub.

On the other hand, we did change constantly the access to coaches, athletes and rest of CAR users. Our aim was to offer better features in a friendly user way.

15.1. Conclusions

In order to measure the success of the experiment, we chose average time session in the website. CARVIREN is a private social network and the time users spend on it has to be correlated with the quality of the content.

We had an average time during the 2 runs of 16.97 minutes.

We realize that CARVIREN is not directly comparable to public social networks, e.g. Facebook or Twitter. We need to remember CARVIREN is a trial of a new environment so people would be bound to be curious about it and thus raise the time spent on it. However, the time spent, together with the feedback received by the athletes, coaches are certainly encouraging indications that this is something of definite interest to the people involved.

Following, we have a report from last January (2014) published by the Wall Street Journal. In this report we can find the minutes per visitor. As we said before, we cannot compare CARVIREN to these social networks, but the following graphic gives us an idea of the current situation:

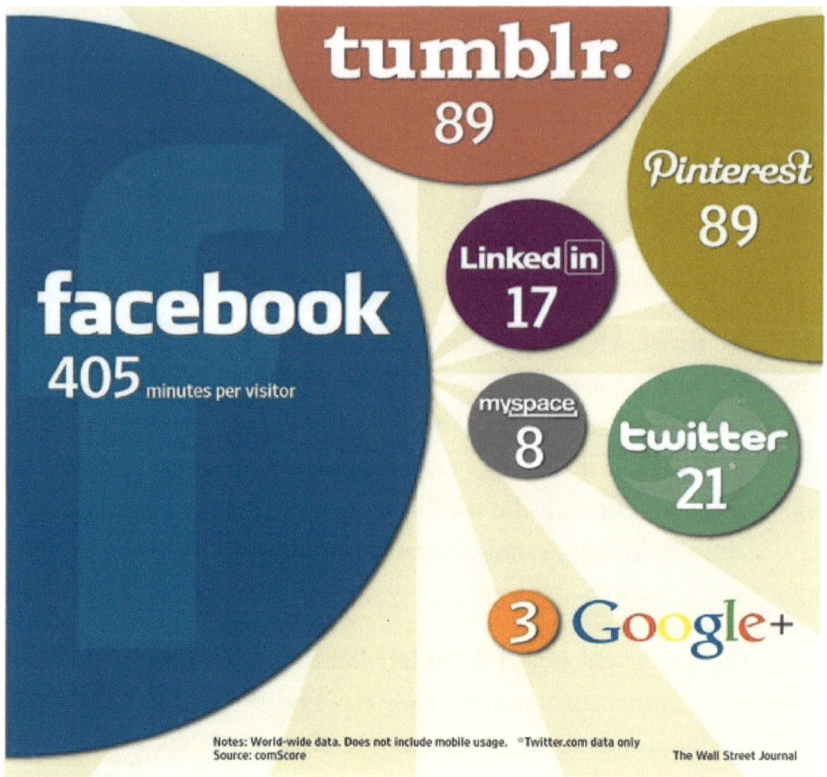

Source: Wall Street Journal

Figure 39: Average time spent in the social network per visitor and month

This time is an average, and it's not the session time, but we can do an extrapolation for different number of visits:

Table 2: Monthly time spent and number of visits for various social networks

			Visits			
	Monthly time (min))	1	15	30	60	90
Facebook	405	405	27,00	13,50	6,75	4,50
Twitter	21	21	1,40	0,70	0,35	0,23
LinkedIn	17	17	1,13	0,57	0,28	0,19
MySpace	8	8	0,53	0,27	0,13	0,09
Tumblr	89	89	6,00	3,00	1,50	1,00
Pinterest	89	89	6,00	3,00	1,50	1,00
CARVIREN	Fixed Value: 17 minutes					

The monthly time spent on social networks is not telling us the number of visits. We can see that engagement on CARVIREN appears good. It would be interesting to know what could happen if CARVIREN continues in the future.

16.   Ethics and Privacy on the Experiment

According to detailed ethical guidelines in EXPERIMEDIA deliverable D5.1.2, all EXPERIMEDIA experiments need to be conducted in accordance with certain ethical oversight procedures. Those principles were integrated in an adequate way into the design of the CARVIREN experiment.

The only update from the last deliverable (4.12.2) was under the **formal requirements**.

CARVIREN experiment intends to record actual motion and sensible data from real athletes in the CAR Venue. This situation was discussed during the EXPERIMEDIA General Assembly held in Madrid in October 2013 and in Leuven in December 2013. It was agreed that we will sign a Controller-Processor Agreement under the current Data Protection Act (DPA) used at CAR.

This document was finalized and signed during the General Assembly in Barcelona, in May, between RTS and CAR.

Before that, no real data was collected.

17.   Conclusions

17.1.   General

Overall the experiment has been successful in many aspects. First of all, the fact that CARVIREN, the social private network for CAR, had a high average time spent in the website. Combined with the positive feedback received by the athletes and their entire environment, it makes us optimistic about a possible commercial acceptance of a future version of this network as a product.

As a future commercial product, almost all parts of the work done will remain.

RTS has been working in high performance for almost 7 years. We are used to developing new monitors (the Sportwis in this experiment). A project like this one, gave us the opportunity to test new areas we haven't tested before (trampolining and oxygen saturation).

Furthermore, we had the possibility to integrate interesting new FMI technologies like the ones of the EXPERIMEDIA Facility, namely AVCC and SCC.

Regarding the Experiment, the most important part was being able to record videos and training sessions, which according to the QoS and the QoE was a moderate-high result. This part was accomplished.

Another important achievement was the synchronization of data with video. This has a great impact on sport training, as when reviewing workout sessions the simultaneous production of data and video can be of great value only if appropriately synchronized.

The sport widgets are also an important achievement. Trampolining air time was not easy at the beginning to calculate, because of the great number of parameters. It required a lot of time and simulations. Final result is a widget with an error no higher than other professional machines.

About CARVIREN's website, there were weaknesses, mostly due to the lack of user friendliness. But as any other platform, improvements came with use and suggestions. Changes have been implemented leading to a better platform.

There were also other areas to achieve improvements. More widgets and more cameras on different rooms could be integrated into a commercial version.

We cannot forget that it will never be as big as a public social network, so potential users are not going to be a large sample, maybe 30-40 users.

EXPERIMEDIA Components have turned out to be great assets. They were reliable and worked as expected and the support for each component was really quick and efficient. As commercial services, the components could be a very competitive solution.

17.2.   Feedback on EXPERIMEDIA Baseline Components

With the CARVIREN experiment, we have focused on using the ECC as a metric logging system, the AVCC for video recording and the Social Annotation Service (SCC-AVCC integration) for the social integration.

The experience has been very good, not only because those are very good tools, but also because we had very good support.

17.2.1. ECC Feedback

ECC has proved to be a very powerful tool. The amount of data and detail is a clear key advantage.

Although it might appear complicated, it is as easy to use as any other monitor tool. In version 2.1, we found an incredible stability and fluency. Also, a very useful feature is the possibility to collect and view live metrics.

Furthermore, the possibility to add more metrics (in our case, our hubs and average usage time) is important. So, you combine metrics from Baseline Components with any other data you need. The fact that it is available for C#, C++, Android and Java is very useful as it allows compatibility with many different systems.

You can export all data in CSV which is much better (when you want to keep control and do your own stats) than XLS (because of the huge amount of data, datasheet cannot responded).

Bottom-line, it's a very powerful tool, very flexible and with many possible applications.

17.2.2. AVCC Feedback

It's a very easy to integrated component, which an incredible support Company behind. Every single time we experimented problems with the video (too slow, fast) we get rapid-response, and we were able to solve all the problems. We haven't used all functionalities, but the few we have implemented on CARVIREN work perfect.

17.2.3. Social Annotation Service Feedback

This was the easiest Component to integrate. It surprises all of us, how simple and easy it was to integrate. Obviously, there is a significant amount of work behind, but the performance during the Experiment was smooth without any problems at all and the service was very user-friendly.

From a commercial point of view, the event on Facebook (combined with the video) is something that can save a lot of time and it's an application customers could look to integrate with their systems.

EXPERIMEDIA
PARTICIPANTS

18. Experimedia Partners

18.1. Core Partners

IT Innovation Centre
http://www.it-innovation.soton.ac.uk/

Ε Π Ι Σ Ε Υ (Institute of Communication and Computer Systems)
www.iccs.gr/

Atos
http://atos.net/

Joanneum Research
www.joanneum.at/.

Infonova
https://www.infonova.com

Foundation of the Helenic World
http://www.ime.gr/

Schladming 2030 GmbH: Startseite
http://www.schladming2030.at/

Centre d'alt rendiment de Sant Cugat del Vallés
http://www.car.edu/es

Katholieke Universiteit Leuven
http://www.kuleuven.be/

Interactive
https://www.tii.se/

Centre for Research and Technology Hellas (CERTH)
http://www.certh.gr/

18.2.   Experiment Open Call 1

IN2 search interfaces development Ltd
http://www.in-two.com/

STI International
http://www.sti2.org/

Graz University of Technology
www.tugraz.at

Public Research Centre Henri Tudor
http://www.tudor.lu/

Universidade de Vigo
www.uvigo.es/

STT
http://www.stt-systems.com/

Poznań Supercomputing and Networking Center (PSNC)
http://www.man.poznan.pl/

18.3.   Experimentes Open Call 2

RealTrack Systems
http://realtracksystems.com

Tecnalia
www.tecnalia.com

Evolaris
www.evolaris.net/

Qualisys
http://www.qualisys.com/

TNO
https://www.tno.nl

REALTRACK SYSTEMS

RealTrack Systems is a Technology-Based Company whose main activity is focused on Sport Science, New Communications and Physical Activity Monitoring. It was founded in 2008 and has developed wireless solutions for Auctions, Portable physical Monitoring Devices and Group workouts monitoring systems, whose main development is a system for monitoring physical activity, the company is specialist in developed hardware and software for sport. The Workforce is a multidisciplinary group made up of highly trained professionals. RealTrack Systems' mission is to provide the athlete, coach, fitness coach, physiotherapist, or team doctor, accurate and relevant information on an ongoing basis on the implementation of the activity. This information is obtained in a non intrusive and can be obtained in the context in which the athlete normally trains. The process improvement training and monitoring of physical activity through the monitoring of certain physiological and kinematic parameters, carries a greater control of the different variables that influence this process, either in the formation processes, high performance or on health. It is therefore necessary to analyze the available technology and its potential use in monitoring training for sport and physical activity.